A
Byrd Anthology

14 Anthems and Motets

Selected by

JOHN MILSOM

Music Department
OXFORD UNIVERSITY PRESS
Oxford and New York

Oxford University Press, Walton Street, Oxford OX2 6DP, England

Oxford University Press Inc., 198 Madison Avenue, New York, NY 10016, USA

Oxford is a trade mark of Oxford University Press

© Oxford University Press 1996

ISBN 0 19 3520079

CONTENTS

PREFACE

Critical opinion today places William Byrd at the very pinnacle of artistic achievement in Tudor music. In doing so, it reflects the view of Elizabethan and Jacobean connoisseurs, who regarded Byrd as their most prized musical possession. Although he was prolific—more music by Byrd survives than by any earlier English composer—he never compromised quality to quantity. There is virtually nothing in his output that is workaday or merely functional.

Byrd was also a versatile composer. He wrote works for choirs, for consorts of solo voices, for viols, for keyboard instruments, and for various combinations of those resources. In his own lifetime, Byrd's music could be heard in cathedrals and churches, in private chapels, at court, in the houses of the nobility, and in domestic amateur music-making. It even rose above faction. Although English society at the time was deeply divided by religious differences, Byrd was nevertheless admired and performed as much by the Protestants as he was by the Roman Catholics with whom he closely allied himself.

Quality, quantity, and variety all make difficult the task of catching the essence of Byrd in one slender volume. This *Byrd Anthology* tries to hint at the riches available, but it frankly admits defeat over the matter of representing Byrd's versatility. Where Tudor music-books could freely mix instrumental works with vocal, and sacred pieces with profane, today's books need to address more specific markets, and this one is no exception. Its aim is to provide modern choirs with a selection that will suit both the church and the concert platform. Missing from this volume are pieces that have secular texts, or call for the use of instruments, or use voice-combinations that no longer fit today's standard choral resources. Also excluded are pieces of considerable length, and of extreme difficulty.

That last criterion calls for some explanation. Many of Byrd's vocal works were composed not with choirs in mind but rather consorts of solo voices. Into that category fall most, perhaps even all, of his Latin-texted motets—or, as Byrd himself called them, 'cantiones sacrae' ('sacred songs')—together with many English-texted pieces that look like church anthems, but which Byrd himself designed for use in the chamber. These 'sacred songs' for solo voices, in both English and Latin, far outnumber Byrd's truly choral works. In manner they tend to be vocally demanding, sometimes even virtuosic. Some of them transfer reasonably well to choral performance. Others, however, are arguably best left to groups of solo singers who can address them, as they would a madrigal, with the sensitivity and agility that their words and music demand. [1]

There are in fact only two authentically choral works in this volume: *Prevent us, O Lord* and *Sing joyfully*, both of which were probably composed for the choir of Queen Elizabeth's Chapel Royal, where Byrd served as a 'gentleman' (or singing member) and organist. The remaining works in this *Byrd Anthology* are all strictly speaking 'sacred songs', borrowed from the chamber repertoire. They fall into three categories: first, *cantiones sacrae* with texts of a general nature; second, *cantiones sacrae* designed to be inserted into the Roman Catholic liturgy; and third, English-texted sacred songs.

Byrd's *cantiones sacrae* on general texts were composed during the 1570s and 1580s. Most were published in books that are dated 1575, 1589, and 1591. An exception is *Audivi vocem de caelo*, which is possibly the earliest piece in this anthology. It owes a clear debt to Tallis, and Byrd himself may have classed it among his juvenilia, for he chose not to publish it; today it survives only in manuscript copies. *Emendemus in melius*, printed in 1575, is modelled on music by Alfonso Ferrabosco, an Italian composer of Byrd's own age who was resident for some years at Queen Elizabeth's court; evidently the two men vied with one another artistically. Also composed probably in the 1570s, though published only in 1589, is *Ne irascaris, Domine*, parts of which borrow from music by Philippe van Wilder, the most prominent musician at the court of Elizabeth's father, King Henry VIII. The three remaining motets in this category, *Domine, secundum multitudinem*, *Domine, salva nos* and *Haec dies* are works of greater independence of thought, and were presumably written relatively close to their publication dates of 1589 and 1591.

No clear pattern emerges in Byrd's choice of text before the 1590s. Some texts derive from the Bible; some are borrowed from motets by foreign composers; a few appear to have been written or compiled by the composer himself, or by his Catholic friends. After 1590, matters change. Byrd now drew his words almost exclusively from Roman Catholic liturgical sources, the Missal and the Breviary. His aim was to provide musical settings for English Catholics to sing during their clandestine services. First to be composed were his three well-known settings of the Mass Ordinary (the Kyrie, Gloria, Credo, Sanctus, and Agnus Dei) for three, four, and five voices respectively. In 1605 and 1607 he published

[1] for further details, see my article 'Sacred Songs in the Chamber', in *English Choral Practice, 1400–1650*, ed. John Morehen (Cambridge 1995), pp. 161–179.

a more ambitious project: two volumes of *Gradualia*, which contain settings of Mass Proper texts (the Introit, Gradual, Alleluia, Offertory and Communion) for some of the main feasts of the Roman calendar, together with miscellaneous pieces on words drawn largely from the Breviary. Three *Gradualia* motets have been chosen for this anthology: *Ego sum panis vivus*, *Justorum animae*, and *O magnum misterium*.

The sacred songs with English texts also date from Byrd's last years. Two of them, *Arise, Lord, into thy rest* and *Praise our Lord, all ye gentiles*, come from his *Psalms, songs, and sonnets* of 1611, the third and last of Byrd's song-books. Three years later he contributed four pieces to a multi-composer volume commissioned by Sir William Leighton, called *The tears or lamentations of a sorrowful soul*. One of them is the song *Come, help, O God*. By this time, Byrd had passed his seventieth birthday. As far as we know, he composed nothing more.

JOHN MILSOM
Oxford, 1996

Arise, Lord, into thy rest

Edited by John Milsom

Psalm 132: 8–9

WILLIAM BYRD
(1543 – 1623)

Source William Byrd, *Psalmes, Songs, and Sonnets* (London, 1611), no. 18.

Printed in Great Britain

OXFORD UNIVERSTY PRESS, MUSIC DEPARTMENT, WALTON STREET, OXFORD OX2 6DP

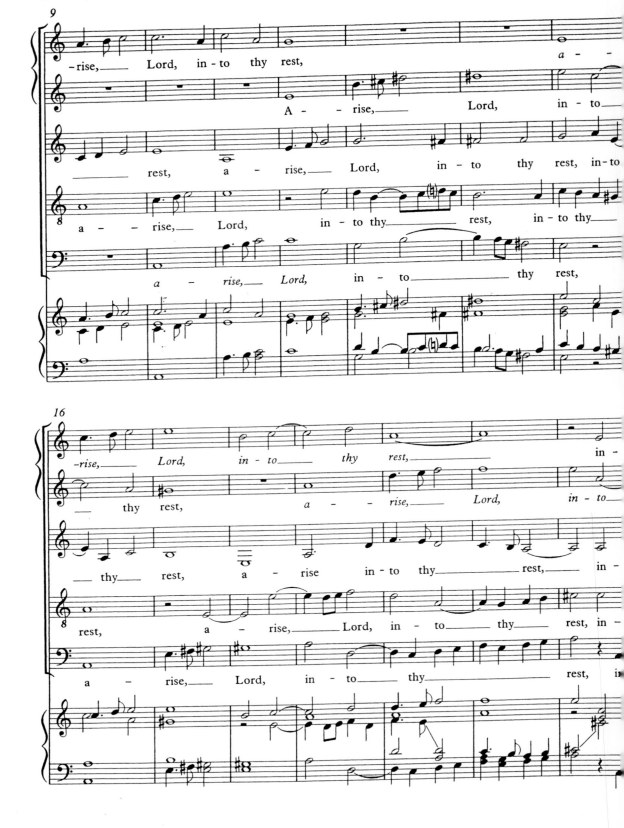

¹ Underlay *sic* in source

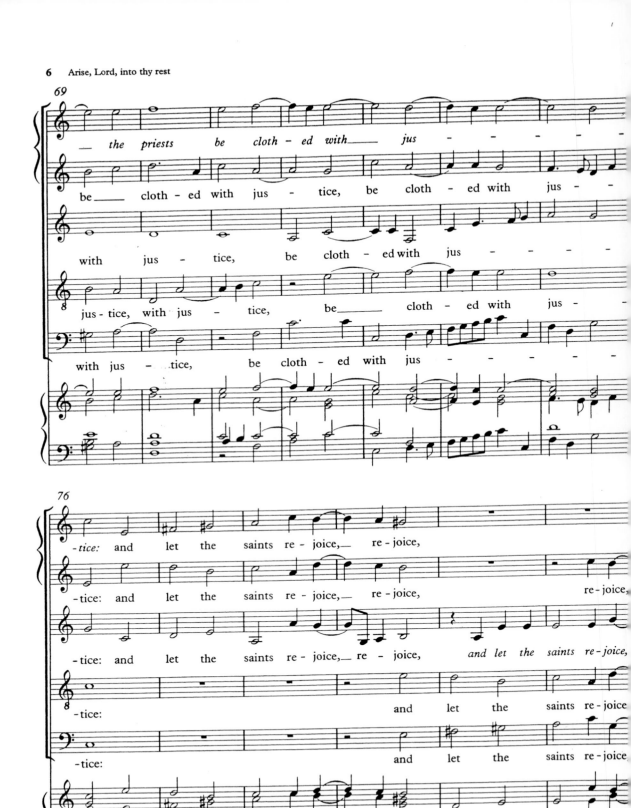

93
re - joice, re - joice, re - joice, re-joice, re - joice, re - joice,
re - joice, re - joice, re - joice, re - joice, and let the saints
-joice, re - joice, re - joice, re - joice, and let the
- joice, re - joice, re - joice, re - joice, re - joice, and let the saints re-joice, re -
re - joice, re - joice, re - joice, re - joice, and let the
98
and let the saints re - joice, re - joice.
re - joice, re - joice, re - joice, re - joice.
saints re - joice, re - joice, and let the saints re - joice.
- joice, and let the saints re - joice, re - joice.
saints re - joice, re - joice.

Audivi vocem de caelo

Edited by Alan Brown

Revelation 14 : 13

WILLIAM BYRD
(1543 – 1623)

[1] 5.3 – 6.4, **A** : no syllables beneath these notes

Translation: I heard a voice from heaven, saying: Blessed are the dead who die in the Lord.

[3] 33.1 – 34.2, **D, E, F:** mor[m] - tu [c. q c c]

Sources

A   Oxford, Christ Church, Mus. 984–8, no. 33 (complete set of five partbooks; hand of Robert Dow, *c*.1581).

B   Oxford, Christ Church, Mus. 979–83, no. 12 (five partbooks from a set of six, lacking Tenor; hand of John Baldwin, late 16th century).

C   Oxford, Bodleian Library, MS Mus. Sch. e. 423, p. 157 (Alto 2 only; late 16th century).

D   Oxford, Bodleian Library, Tenbury MS 389, p. 141, and MS in the private possession of Mr David McGhie (the 'James' MS), p. 134 (Alto 1 and Soprano respectively; late 16th century).

Sources **E** to **H** are from the Paston collection and may be dated *c*.1600–1610.

E   Oxford, Bodleian Library, Tenbury MSS 341–4, f. 11v etc. (four partbooks from a set of five, lacking Bass).

F   Oxford, Bodleian Library, Tenbury MSS 369–73, f. 25v (complete set of five partbooks).

G   Chelmsford, Essex County Record Office, MS D/DP.Z6/1, f. 28 (Bass only).

H   London, British Library, Add. MS 29247, f. 15v, and London, Royal College of Music, MS 2089, f. 55v (identical copies of lute arr. of voices II–V, a tone above original pitch).

I   London, British Library, Add. 5058, p. 73 (18th-century score, not collated).

Critical Commentary

T, 12.2, **A**: no accidental / B, 14.2, **B**: no accidental / A2, 27.3–28.1, **A**: slurred / B, 39.2, **A**: E for C / The following are slurred in **A**, in each case resulting in the syllables '-ri-un' being placed later: A1, 41.2–3; T, 47.2–3; B, 50.4–51.3; A1, 55.3–56.2; T, 61.2–3 / A1, 50.2, **A, D**: no accidental / A2, 55.2, **B**: no accidental / A1, 60–62, **A**: Doᶜ-miꟼ-noᶜmoᶜ ᶜ-riꟼ-unᶜ-tur,ᶜ moᶜ-riꟼ-unᶜ ᶜ / T, 63.4–65.2, **A**: Doᶜ-miꟼ-noᶜ moᶜ- riᶜ-unᶜ ᶜ / S, 67.4–68.2, **B**: mC

Come, help, O God

Edited by John Milsom
Sir William Leighton (*c.*1565–1622)

WILLIAM BYRD
(1543 – 1623)

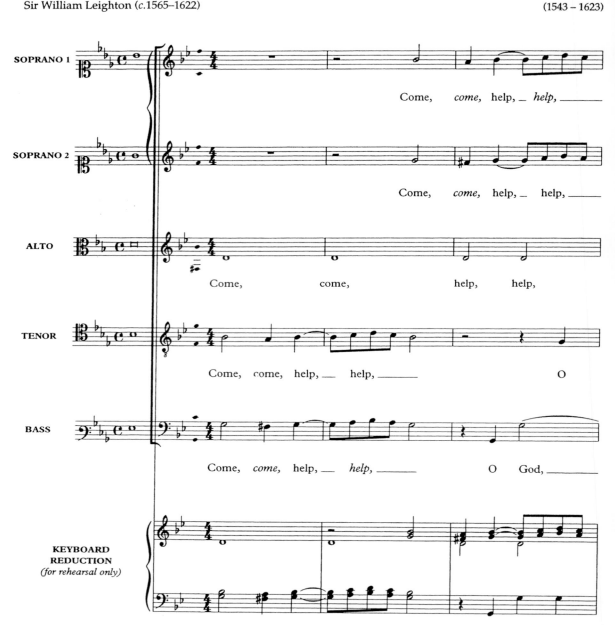

Source: *The Teares or Lamentations of a Sorrowfull Soule* (London, 1614)

Domine, salva nos

Edited by Timothy Symons

Magnificat Antiphon, fourth Sunday after Epiphany (Roman rite)

WILLIAM BYRD
(1543 – 1623)

[1] '- va' placed here

Source: William Byrd, *Liber secundus sacrarum cantionum* (London, 1591), no. 31.

Translation: Save us, O Lord, for we perish: command and create peace, O God.

Domine, secundum multitudinem

Edited by Timothy Symons

Psalm 93 (94) : 19

WILLIAM BYRD
(1543 – 1623)

Source: William Byrd, *Liber primus sacrarum cantionum* (London, 1589), no. 27.

Translation: O Lord, according to the multitude of the sorrows in my heart, thy consolations have gladdened my soul.

Ego sum panis vivus

Edited by David Skinner

Antiphon to the Benedictus at Lauds
for the feast of Corpus Christi.

WILLIAM BYRD
(1543 – 1623)

Source : William Byrd, *Gradualia . . . lib. secundus* (London, 1607; reissued 1610), no. 17.

Translation : I am the living bread, which ascended into heaven: if anyone eats of this bread, he will have eternal life. Alleluia.

Emendemus in melius

Edited by Timothy Symons

2nd Respond at Mass, Ash Wednesday (Roman rite)

WILLIAM BYR[
(1543 – 162]

Source : William Byrd and Thomas Tallis, *Cantiones . . . sacrae* (London, 1575), no. 4.

Translation : *i)* Let us atone for the sins we have committed in our ignorance, lest, should the day of death suddenly overtake us, we seek time for repentance and cannot find it. Hearken, O Lord, and have mercy, for we have sinned against thee.
ii) Help us, O God of our salvation, and, according to the honour of thy name, deliver us.

SECUNDA PARS

Haec dies

Edited by Peter le Huray
and David Willcocks

WILLIAM BYRD
(1543 –1623)

Psalm 118: 24

Source: *Liber Secundas Sacrarum Cantionum* (London, 1591), No. XXXII.

Translation: This is the day which the Lord hath made; we will rejoice and be glad in it. Alleluia.

5
- nus, haec di - - - es quam fe-cit Do - mi-nus, haec
di - - - es quam fe-cit Do - mi - nus,
- nus, quam fe-cit Do - mi - nus, Do - - - mi -
Haec di - -
Haec di - - - es quam fe-cit Do - mi-nus, quam
Haec di - - -
10
di - - - es quam fe-cit Do - - mi - nus,
haec di - - - es quam fe - cit Do - - -
- nus, quam fe-cit Do - mi-nus, haec di - -
- - es quam fe - cit Do - mi-nus, Do - - mi - nus, haec di - - -
fe - cit Do - - - - mi-nus, haec di - - - es
- es quam fe - cit Do - mi-nus, quam fe - cit Do - mi-nus, haec

28
et lae-te - mur in e - - - - - a. Ex - ul - te - mus,
Ex - ul - te - mus
et lae-te - mur, et lae-te - mur in e - a. Ex-ul-te - mus, ex-ul-te -
et lae-te - mur, et lae-te - mur in e - a. Ex - ul -
et lae-te - mur in e - a. Ex-ul-te - mus, ex-ul-te - mus
et lae-te - mur. Ex - ul - te - mus,
34
ex - ul - te - mus et lae-te - mur in e - a,
et lae-te - mur, et lae-te - mur,
- mus, ex-ul-te - mus et lae-te - mur, et lae-te - mur, et lae-te -
- te - mus et lae-te - mur, et lae-te mur, et lae-te -
et lae-te - mur, et lae-te - mur, et lae-te - mur, et lae-te - mur,
ex - ul - te - mus et lae-te - mur,

Justorum animae

Edited by Anthony Greening

WILLIAM BYRD
(1543 –1623)

Offertory for All Saints' Day
Wisdom of Solomon 3, 1–3

Source: William Byrd, *Gradualia ac cantiones sacrae* (London, 1605)

Translation: The souls of the righteous are in the hand of God, and there no torment shall touch them.
In the sight of the unwise they seemed to die: but they are in peace.

Ne irascaris

Edited by Timothy Symons

Isaiah 64: 9–10

WILLIAM BYRD
(1543 – 1623)

Translation: Be not wroth very sore, O Lord, neither remember iniquity for ever: behold, see, we beseech thee, we are all thy people.

Thy holy cities are a wilderness, Zion is a wilderness, Jerusalem a desolation.

⁹ A: E sharp, presumably in error   ¹⁰ A: m, m-rest

[11] A: '-ta' possibly placed here   [12] B, C, D: mm for m.c

13 B: m for cc, '-ta' to rest of bar 130, 'est' to 131   14 B, C, D: '-so-' placed here   15 D: '-la-' placed here

[17] B: '-ta' placed here   [18] A: '-ta' placed here   [19] C: cc for m; '-ta' placed under second c

Sources

A   William Byrd, *Liber primus sacrarum cantionum* (London, 1589), nos. 20–21

B   Oxford, Bodleian Library, MSS Mus. Sch. e. 1–5 (1580s)

C   Oxford, Christ Church, Mus. 984–8, no. 9 (1580s)

D   Oxford, Christ Church, Mus. 979–83, no. 59 (1580s)

Other sources not consulted for this edition are listed in *Cantiones Sacrae 1589*, ed. Alan Brown, *The Byrd Edition*, vol. 2 (London, 1988), which includes a full listing of variant readings in all sources.

Commentary

This edition is based on **A**, which is likely to have been supervised by the composer. Points where the text-underlay of **A** is ambiguous or probably misplaced are indicated by footnotes. In a few cases the readings of sources **B**, **C**, and **D** have been preferred and are incorporated into the edition; elsewhere they are offered in footnotes as alternatives.

O magnum misterium

Edited by David Skinner

Matins respond, Christmas Day

WILLIAM BYRD
(1543 – 1623)

Source: William Byrd, *Gradualia . . . lib. secundus* (London, 1607; reissued 1610), nos. 8 – 9.

Translation: O great and wonderful sacrament, that beasts should see the birth of our Lord, lying in a stable. O blessed Virgin, whose womb was worthy to bear Christ our Lord. Hail Mary, full of grace, the Lord is with thee.

Repeat 'Beata Virgo', bars 45–61.

Praise our Lord, all ye Gentiles

Edited by Timothy Symons

Psalm 116 (117)

WILLIAM BYRD
(1543 – 1623)

Source : William Byrd, *Psalmes, Songs and Sonnets* (London, 1611), no. 29.

[1] original underlay '- main ^{c.q}-eth ^cfor ^cev ^m-'

2 C for E

Prevent us, O Lord

Edited by David Skinner

WILLIAM BYRD
(1543 – 1623)

A prayer

Sources

Byrd's *Prevent us, O Lord* survives in no fewer than 27 manuscript sources, as well as John Barnard's *The First Book of Selected Church Musick* (London, 1641). The present edition is based on Oxford, Christ Church, Mus. 984–988, no. '59' (correctly no. 58), a complete set of five partbooks copied in 1580s, which is the earliest surviving source, and also the most accurate. The Commentary below discusses issues arising from this source. For a full critical report, see *The English Anthems*, ed. Craig Monson, *The Byrd Edition*, vol. 11 (London, 1983), no. 8.

Commentary

Bar 18, Alto 1:  ; the edition follows the reading of other sources, which avoids consecutive fifths.

Bars 18–19, Bass: 'help' ambiguously placed.

Bar 33, Bass:  ; changed to agree with Tenor.

Sing Joyfully

Edited by John Morehen

Psalm 81:1–4 (Geneva Bible)

WILLIAM BYRD
(1543–1623)

Editorial Note

This anthem, one of the most popular of its period, survives in about a hundred manuscript and printed sources of the early seventeenth century; no sixteenth-century sources are extant. No overall stemmatic relationship of the sources can be established, and this edition mainly represents the results of a collation of the earliest sources with those later sources that are of proven authority for music by Chapel Royal composers. For a complete listing of both sources and variants, see William Byrd, *The English Anthems,* ed. Craig Monson, *The Byrd Edition,* Vol. 11 (London, 1983), no. 10. The organ part in the present edition is taken from Durham Cathedral Library, MS A1, p. 167 (*c.*1635), and may be omitted in performance.

12
loud, sing loud un - to
un - to the God of Ja - - - - cob, un -
- to the God of Ja - cob, sing loud un -
loud un - to the God of Ja - cob, sing loud
un - to the God of Ja - - - - cob,
loud un - to the God of Ja - cob, un -
15
the God of Ja - cob.
- to the God of Ja - - - cob.
- to the God of Ja - cob. Take the song,
un - to the God of Ja - - cob. Take the song,
the God of Ja - cob. Take the song,
- to the God of Ja - cob.

25
the vi - - ol, the pleasant
the pleasant harp and the vi - ol, and the
harp and the vi - ol, the plea - sant harp, the
the pleasant harp and the vi - ol, the vi -
and the vi - ol, the plea - sant harp,
plea - sant harp, the pleasant harp and the vi - ol, and the vi - ol,
28
harp and the vi - ol, and the vi - - ol.
vi - ol. Blow the
pleasant harp and the vi - - - ol. Blow the
- ol, and the vi - - - - ol.
plea - sant harp and the vi - ol. Blow the
and the vi - - ol. Blow the

50
day. For this is a sta - tute for Is -
day. For this is a sta - tute for
feast day. For this is a sta - tute for
day. For this is a sta - - - tute
feast day. For this, for this is a sta - tute
feast day. For this is a sta - tute for
54
- - - ra - el, and a law of the God of Ja - cob, God
Is - ra - el, of the God
Is - ra - el, and a law of the God of Ja - cob,
for Is - ra - el,
for Is - ra - el, and a law of the
Is - ra - el, and a law of the God of Ja - cob, of

Processed, Set and Printed by
Halstan & Co. Ltd., Amersham, Bucks., England